UNIVERSAL EXPANSION AND BLACK HOLES

Printed in the United States by Eastman Technologies

First Printing, 2016

ISBN: 1541288939

DEDICATION

This book's dedicated to my late grandfather, Arthur "Lee" Eastman; the smartest man I've ever known. He began inspiring me at a young age to enjoy electronics and science with amateur radio, aviation and computers.

I'll forever be thankful for him purchasing my very first computer, which was a Commodore VIC-20, along with a matching cassette tape drive.

ACKNOWLEDGMENTS

It's imperative to thank Adam Wimbish, who knows more about physics than anyone I know, and, who believed in me when many others declared I was going mad after hearing my "universal expansion" theory. Adam helped me develop and flush out my equations and gave me much feedback, all of which I'm deeply thankful.

CONTENTS

INTRODUCTION

Nobody sits down with the purpose of writing a theory (let alone a book) which attempts to discern why the universe is expanding. I rather preferred to simply publish this work as an academic paper, but lack the standing, as I'm not faculty or a graduate student in college. Not many academic journals are thrilled about publishing papers by authors whose experience is generally outside the field, or isn't a faculty or graduate student. Upon discovering and working out the problem of universal expansion, the need for its proper documentation became evident. The circular hurdle mentioned above predicated this thrilling work would be prepared as a book instead.

After several years of explaining extremely technical consumer electronics problems to audiences without the required years of acumen and background necessary to understand the subject matter, I was thrilled to have the chance to also reach those outside the academic physics community. This would've been difficult to achieve if published in an academic journal; it'd have long equations on every page, which simply are expressing (in a very complicated manner) the factual descriptions of the claims and principals. This causes much important work in physics to be somewhat "out of reach" for the casual reader, even if they've had some formal background in a few secondary school or college courses. Many people without physics backgrounds are interested in both black holes themselves, and, the vexing problem of why the universe's expanding faster; which nobody in the world's solved...until now.

Black holes are inherently at the root of why our universe's not only expanding, but doing so at an accelerated rate. Others have explained the dynamics of how black holes work in space-time, but the important step of linking the two with empirical formulas that account for the universes expansion (as a direct result of black holes' behavior) has not been achieved. That's the goal of this text; to properly link the behavior of both in a cohesive and responsible manner that both academics and casual readers can appreciate and find useful.

My original aim was never to unearth such a unifying theory; in the same manner that many useful inventions are generated solely as the result of solving a focused problem and not as the result of trying to intentionally generate a grandiose new product or technology. One of my favorite physicists, Stephen Hawking, long ago advanced the theory of *Hawking radiation*; [1] which I've never been able to properly believe; despite being impressed at the far bigger problem of "information loss" with black holes that he was trying to solve. Few people have tried to work on that problem with any resounding success and its instructive to give him much credit for tackling it. I've always thought it suspect, however, that any object with so much gravitational force as a black hole (that even light couldn't escape the powerful grip of) would suddenly start emitting radiation; regardless of the other dynamics of space–time which are inherently involved. Quasars are far chattier than black holes and spew cosmic noise out everywhere into the vacuum of space; however, black holes are more low-key in a sinister way; everything which gets near them simply disappears, regardless of whether it's matter that can even be observed.

Nothing claiming even the most superior escape velocity in the universe would simply ooze out a radioactive mess for others to detect; a black hole derives its very name for being the most powerful, yet hardest to observe phenomena in the galaxy…doesn't it? A living organism in natural science adapts and evolves to become invisible (or difficult to spot) if it lacks the capability to defend itself from others in its environment with any reasonable offensive efficiency. Celestial objects age and change states constantly, with gravity generally being the capability making them superior to others during their observable lifecycle. Generally, if you possess stronger gravitation than your neighbors, you're the shark in the astrological waters and exist at the top of the food chain. If you have significantly less gravitation, you're an injured fish that will fit in just about everyone else's mouth. Black holes have the added edge of not only being the very largest sharks, but also being completely invisible. Not only is a black hole going to always win any gravity contest, the objects being consumed by one will never even "see it" coming.

I've long maintained with colleagues, for example, that if such an event as *Hawking radiation* was occurring, it'd be happening on "the other side" of the black hole's singularity, meaning that it'd also be undetectable or visible to observers on the outside of the singularity. Even so, it'd still make little sense that the most powerful objects in the universe would effectively "burp" and continually hold all that other gas inside its stomach while also eating, nonstop, all the time. Nothing which exists in the natural sciences can do so either; even bears and sharks cannot eat constantly, and, even if they could, they couldn't do so without passing gas and being sick. In merely trying to explain why it was absurd (if only in my own mind as a thought experiment) for black holes to emit *Hawking radiation*, I stumbled upon the more perplexing solution for explaining why the universe itself was expanding, and, why it was expanding faster; which contemporary physics has been so far unable to answer conclusively, outside of some interesting experiments with redshift measurements which seem to prove only that the expansion rate of the universe appears is accelerating. This was not the problem I set out to solve, however, it became an unintentional side effect from my contemplation on universal expansion; my notion of why *Hawking radiation* couldn't occur.

Some contemporary prescription drugs are "discovered" as a side effect completely unrelated to the original research & development goals; it's in this fashion I "accidentally" discovered a method to prove both why the universe was expanding, and, why it was happening faster over time. A lot of people have figured out answers to one problem, but not both, at least not conclusively anyway; since they're inextricably linked, it's necessary to have a credible explanation for both that can then be validated by others' own tests. To make things more frustrating, several good measurements of the expansion rate of the universe have been posited which seem correct; however, they still don't explain why it's occurring. The byproduct of my answer also explains (somewhat) the non-existence of the quantum radiation effect Stephen Hawking has been championing for three decades as being emitted from black holes. In the end, this very complicated problem can be broken down with a very easy solution; albeit using some "prongs" from both *quantum mechanics* and special relativity, which in-themselves are inherently complicated, but easily explained.

I told Adam my *universal expansion theory* and we then discussed it at great length several times; eventually reaching the conclusion that if the answer was both correct and simple, then it should also be expressible with an equation. Merely developing a compelling, normative idea to solve such a difficult problem may be correct, but such reasoning needs empirical scrutiny to be validated. Something as fundamentally important to physics as the expansion rate of the universe being properly defined, I decided, deserved its own empirical foundation (and rules) validated from necessary principals in both standard and quantum physics; from which others, may judge if the theory contained herein is credible, useful, and, able to undergo additional validation in the future. It's sincerely hoped this work succeeds in doing so.

CHAPTER 1

WHAT'S UNIVERSAL EXPANSION?

Universal expansion is, quite simply, a definitive explanation and theory for solving the longstanding problem in astrophysics of why the universe is expanding, why it's doing so at an increasing rate, and finally, why our observations from Earth do not seem to reveal the entire story. The expansion of the universe is inextricably linked to the phenomenon of black holes, and, the initial "surge" that's widely attributed with the beginning event of the universe itself. Regardless of your belief governing how the universe began (some theories hold it's always existed, or, is simply a virtualized series of programs running on computers somewhere) it's accepted by everyone that now on Earth, the universe is steadily expanding away from us at an increasing rate.

Most physicists agree that the universe has been expanding since its initial inception, regardless of their independent cosmological theory as to the actual process occurring that's causing it, or, even how/why the formation of the universe occurred. The advent of observatories, radio telescopes and powerful orbital satellites has proven empirically that the universe is expanding by comparing "snapshots" taken of a given area or body against time and several other methods. In the same vein that observing the curvature of the Earth reduced the theory of the planet being flat in an empirical manner, so too has our celestial observations with respect to the expansion of the universe.

Physics has so far well described empirically both the phenomenon of the universe expanding (since its initial creation, or "Big Bang" event) and the fact that it's been steadily increasing ever since and continues to do so. Several theories have been advanced which propose that the universe will continue expanding indefinitely and they're correct. The principal hurdle remaining to having a more complete overall understanding of this phenomenon is the simple question; again, of why is the universe expanding? There's been no suitable explanation derived for this problem in which any sound empirical basis accommodating it in astrophysics. Universal expansion, however, achieves both important goals; explaining not only why the universe is expanding, but also why it's increasing, and finally, why it will continue to infinitely increase; or, until the universe somehow ceases to exist.

The inextricable element to solving and understanding the expansion of the universe are black holes. Black holes consume all the matter around them in a circular fashion from a boundary point around them known as an event horizon. The fact that matter cannot escape the intensive gravitational pull of a black hole, and, that even light itself cannot escape is commonly known outside the physics community. There's lesser known problems which have continued since the 1970s in the physics community, dealing with subjects such as "information loss" and the eventual *information paradox theory*, which supposedly occurs as a byproduct of black holes' existence; with the debate of whether matter can partially escape (or copy itself) from the singularity of a black hole still raging today, mostly as the result of Gerard t'Hooft, Leonard Susskind and Stephen Hawking, amongst many others.

The truth, per the *theory of universal expansion*, is that no information is ever lost which enters a black hole; it's simply displaced once it enters the singularity and is stretching the "fabric" itself of space because of time dilation that's unobservable from the outside of the black hole looking in; which is the only method in which we can observe black holes. A more through explanation of the role of time dilation and its effect will be presented later in chapter four. As the universe expands, there are more celestial bodies created which then die at some eventual point, regardless of how many years that process may take.

11

A larger number of ever increasing stars means that the small number of them that will become a black hole, continues to increase in a linear method that's always increasing. As the universe expands, there's more room for new celestial bodies on top of the "fabric" for the already existing ones, which are constantly aging. This effect is essentially an exponential increase in the number of black holes that are being created, and, will continue being created in the future.

Once information enters a black hole, it may be traveling at various speeds, however, most calculations suggest that even light moving at its maximum speed in a vacuum cannot overcome the intensive gravitational pull of a black hole; hence the a traditional notion in physics (which stems directly from Einstein's *theory of general relativity*) that a "cap" exists at the speed of light; with 186,000 miles-per-second being the maximum threshold information can maintain when traveling though the singularity of a black hole. Since the fastest known particles in the universe are obviously light, the crutch that's proverbially been leaned on so heavily by the physics community (in explaining this phenomenon) is the classic proof that light itself can bend, which we have Einstein's *special theory of relativity* to point at authoritatively. Light can and most certainly does bend, however, light bending as it enters the singularity of a black hole simply because it's round is an irresponsible theory that's incomplete and has thus halted further research into the purpose for our ever-expanding universe, which universal expansion so elegantly defines and describes empirically.

As light travels at 186,000 miles-per-second and approaches the singularity of a black hole, it bends because of the intensive gravitational force. While information (or other types of matter besides light, basically) may not be traveling as fast as light, they're still consumed by the singularity of the black hole and "enter" at a rate slower than the speed of light. When light approaches the singularity, however, something different occurs. To physically absorb the light, it first bends because of the circular nature of the singularity in a black hole. Secondly, the lights velocity accelerates because of the gravitational intensity unique to black holes.

The rate of acceleration above the theoretical maximum of 186,000 miles-per-second is achieved when light is consumed by a singularity. This phenomenon can be viewed practically as a spacecraft trying to escape the Earth's gravity and requiring a "greater than" speed value than the "constant" value that describes the Earth's gravity. To leave our planet, one must achieve a speed of 25,000 miles-per-hour, or 40,320 kilometers-per-hour. If the speed of light was instead the escape velocity of our planet and a spacecraft obtained 24,000 miles-per-hour, it would fail to leave the upper reaches of the atmosphere and eventually fall back to the ground. If the normal, zippy speed of light were instead reduced to less than 25,000 miles-per-hour, it too would also fail to escape the gravitational force of Earth. One must reliably travel 25,000 miles-per-hour (or faster) to achieve orbit around Earth because of gravity; to instead escape the gravity of a black hole, one must reliably travel faster than 186,000 miles-per-second.

If the gravitational force here on Earth instead required 186,000 miles-per-second to escape into space and achieve orbit (and the light from the most powerful source available escaped when pointed upwards from the planet's surface, then the only reasonable explanation would be that the gravitation force was simply so uncommonly powerful that it caused the light to develop a speed in-excess of 186,000 miles-per-second. This is because of the fundamentals of inertia; matter traveling in a singular direction always continues to do so, unless an external (or otherwise outside) force is presented. *Newton's first law of motion* generally allows for this phenomenon and cares not what the speed of the objects in question are traveling at, or more broadly even, what dimension of space they may occupy at the moment. *Newton's second law of motion* provides a more empirical basis for working such matters out; the familiar **F = ma** equation for calculating force versus mass and acceleration, states that a given mass with a constant velocity will always maintain that velocity, except for cases where an external force effects it, generally causing acceleration. Newton didn't know of the existence of black holes, but nonetheless, his formula still applies the same to objects being affected by a massive external gravitation, or similar force, as is characteristic of a black hole. If light itself cannot escape the immense pull of a black hole at its maximum velocity, the external force it exhibits upon it must cause it to increase in speed as consumed. Einstein allowed for such peculiarities with special relativity and much of the final two chapters are devoted to this topic.

Newton's third law of motion is also perhaps the most quoted in science classrooms and states that, "for every action, there's an equal and opposite reaction." This comprises much of the reason why we cannot "see" inside a black hole and observe the information being pulled into it from a safe distance outside. The opposite reaction in this illustration is proportional to the speed of light increasing past 186,000 miles-per-second, while being consumed by the massive force of the singularity, whereas the initial action is simply light passing "the point of no return" near a black hole. Time dilation is responsible for some aspect of the opposite reaction in this example and causes a static image to appear that we always see representing a black hole with a telescope, however, on the other end of the singularity (and arguably at some point whilst traveling through it during acceleration beyond the speed of light) the same image appears translucent because the solar mass itself has now begun to travel backwards in time, despite being "right in front of us" from a practical perspective.

Given our fixed third dimension in space-time, it's difficult to see something that's partially (or completely) further ahead of us in the same place. The easiest way to visualize this seeming paradox is to have two doors with large glass windows in them situated against each other so that one may swing open one direction and another that also opens outwards the other direction on the reverse side. The information that's being consumed through the glass window in the door facing Earth can be seen, and, the matter appears to endlessly disappear through the window in a circular pattern; not unlike a giant vacuum cleaner consuming all the dirt from a rug when pushed along the floor. Once we open the first door, however, we have two distinct problems that quickly become noticeable.

First, we cannot see the second door itself, or it's glass window inside, despite being confident that the first door has been opened and that we know it's there from earlier inspection in our own space-time. Secondly, all the light (and other matter) which are being consumed through the singularity of the first open door is no longer visible, or has traces of bent light on the very edges of the door, in a circular vortex pattern. Any light, which had passed into the threshold area of the second door, would not be visible to us because it was ahead of us in time, regardless of whether the measure was in seconds, minutes, or days, for example. Any light that had only just started to eclipse it's generally agreed upon acceleration constant in the vacuum of space of 186,000 miles-per-second would sporadically appear as described above, typically streaking light moving towards the inside of a concentric circle.

The key element to having a mostly invisible second door that's just black when the first one has been opened is congruent not only with Newton's three laws of motion, but also with the rules of both general and special relativity by Albert Einstein. It also explains why we cannot often get a decent visual representation of a black hole without a radio telescope and a lot of observation of the nearest periphery area over time, which reveals drag or pull being caused by the intensive gravitation force. When matter approaches the speed of light the effect of time slowing becomes both apparent and measurable. Whilst mankind hasn't begun traveling at speeds anywhere near that of light yet, we know with certainty that a linear time change occurs that's directly proportional to both the speed being traveled and the time elapsed.

A person who traveled for a year at the speed of light and returned to the planet would have aged consistently less and would effectively be "behind" where they should be at that moment, without the long journey at such high, prolonged speed. We cannot see the inside of a black hole from safely outside the event horizon. This is also why one cannot see that which is moving backwards very quickly; while the entire rest of us and our planet is moving forward at such a slow and generally non-sporadic rate. The second door of the black hole can also be only seen once ones passed the point of no return inside the singularity of a black hole, however, the vast amount of visible information entering it may vary between long curving streaks of light and simply nothing but blackness, for example. Our eyes aren't having fun with us at our own cognitions expense; from our dimension and space-time, it's perfectly acceptable to see nothing. This more thoroughly answers the question of why black holes appear black (and as absolutely nothing to us optically) from astronomical observations on Earth, or near Earth orbit. The effect of time dilation causing this phenomena (with information being consumed and then no longer being visible to us) will be explained in more detail in chapter four, whereas the concept of light accelerating past 186,000 miles-per-second will be discussed in the final two chapters; which addresses the applicability of Einstein's *theory of special relativity*, and, which still explains today how such special scenarios may occur that appear to defy conventional theory, as well as some related observations which have been carried out on Earth, or, in close orbit. Einstein didn't originally have a lot of experiment options available in the early twentieth century to validate his theory that light could bend when he published it; he was simply working through thought exercises to their logical conclusion.

From studying his work, it seems Einstein was trying more to explain "why" such strange phenomenon could occur within the traditional norms of space, than "how" they manage to do so, although he succeeded at both; as times passed, many experiments have been developed which illustrate or empirically prove their existence. Einstein didn't violate any of Newton's three laws of motion and it doesn't appear that this author is doing so either; but rather instead, using them to reinforce concepts in space-time that explain anomalous behavior that appears to divert from the standard model of physics. A black hole is not a common celestial body and while there will always continue to be more of them forming over time because of universal expansion, they represent a very small minority of overall objects in space; they simply have a very important behavior which governs how they necessarily exist. While we cannot visit the singularity of a black hole (even with a probe or satellite) to observe and validate such theories from our perspective on Earth, we can watch solar mass accumulation and the consumption of neighboring matter over time. The standard model of physics is very useful for defining observations, which have no special characteristics inherent with special relativity, for example, however, the constants are often viewed as never being subject to change under any circumstances. In the case of universal expansion and black holes, its necessary to view them less as a rare element of the standard model of physics, and more a case of special relativity and time dilation amongst a very rare object that still observes the same conventions, albeit with different conclusions, methods and reasons. This "standard model" hurdle has been a fundamental stopping block for many talented physicists who might well have otherwise developed this theory themselves many years earlier. We'll discuss this "black hole perspective" problem with competing *standard* and *quantum mechanics* in the next chapter.

CHAPTER 2

WHY THE UNIVERSE IS EXPANDING

In 1921, Sir Oliver Lodge of *Birmingham University* first advanced the theory that an intensive concentration of matter resulted when a star imploded its own solar mass into an extremely tiny size.[2] He attributed this to light being subject to the same rules as gravity and therefore necessarily possessing at least some bit of weight, that could then be retained and prevented from escaping; thus, the massive implosion which begins the life of a *Schwarzschild radius*. The *Schwarzschild radius*[3] is imperative for the study of black holes and physics because it serves as a "point of no return" constant after the creation of a black hole; with matter collapsing inside itself faster and faster until light can no longer escape.

Returning briefly to the example in chapter one about escape velocity, we can calculate that Earth requires exactly 11.2 kilometers-per-second to escape into orbit during normal, everyday conditions. If we were to compress Earth down to a quarter of its size, yet still allow it to retain its original mass, we would quickly find that an increased speed of 22.4 kilometers-per-second was now required to escape our planets gravity. *Newtonian* motion laws are directly applicable to escape velocity and thus relevant to time dilation problems, especially with black holes from our perspective on Earth. *Newton's third law of motion* states that for every action, there's an equal and opposite reaction, one example is two objects coming together, irrespective of speed. The force from the first object equals the size of the force on the second object, or more plainly, whenever an object pushes another, it'll receive an equal amount of force pushing against it. Newton wasn't aware of time dilation or *special relativity theory* yet, however, this law of motion continues to be just as relevant today. Several factors affect the execution of this third law, especially with a black hole, as inherent time dilation occurs when traveling past the singularity. The curved space-time of special relativity and universal expansion together (particularly with respect to black holes) produces an effect in which the speed of light is exceeded when information is consumed by a black hole; thus, time dilation occurs and the event can no longer be viewed by us; because it's now occurring at some point in the past, respective to its approximate velocity. Such cases of this phenomenon (other than just with black holes per se) are important to space-time interval calculations in physics, as they're known more specifically as light cones. Light cones represent a useful way of both charting and visualizing the "progress" of a light beam as its time changes from the present into the past or future, for example. A boundary always exists for both the past (or current time) and the future that's occurring among the same path of light. This helps immensely to both explain and prepare empirical expressions for the progress of this light as it travels between times in space. The observer is always situated between the light cones in such illustrations, because one cannot ever be in two light cones at one time, or, see into more than one of them; these two rules are what govern light cones behavior as expressed by also considering *quantum mechanics.* The age-old phrase about it "being impossible to be in two places at once" applies perfectly to light cones in the sense that even if you're situated rather close to the time division itself, its still impossible to see into a light cone other than the one you're currently in.

The illustration [4] on the following page demonstrates these concepts in a way that's easier to visualize.

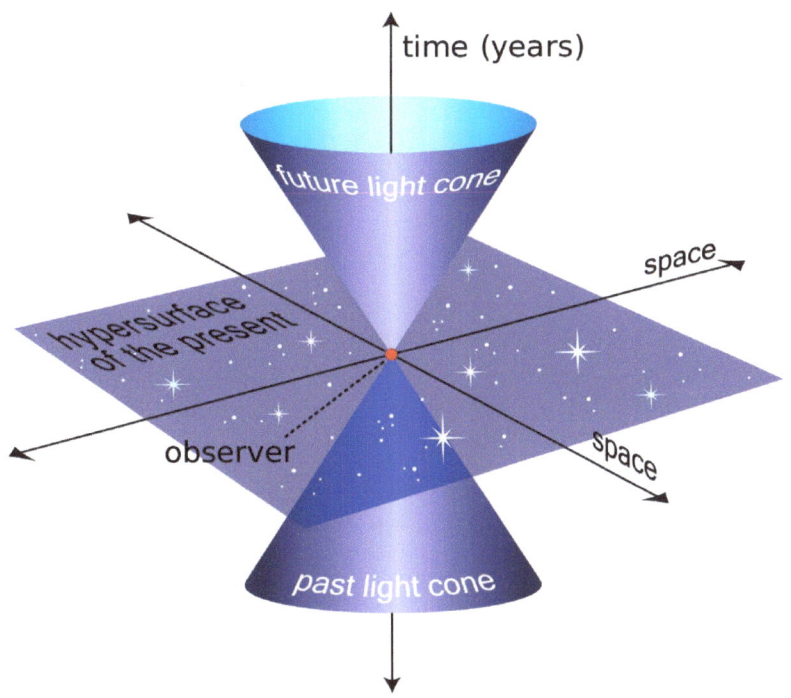

Gravitational lensing is caused in the cases of time dilation problems when entering a singularity; the boundary for the past and future no longer appear equal. In other words, the light cone isn't straight anymore and has been curved so that part of it exists only in the future and it no longer shares a boundary with the current time, or past. In the case of universal expansion, we can view a light cone from our perspective on Earth traveling towards a black hole, however, we cannot ever see inside or "past" it sideways, as those areas are only occurring in the past of the light cone and are impossible to see.

The easiest way to envision this is imagining that you were in the future light cone "inside" of a black hole and waiting for a colleague to arrive for lunch who's traveling on a light cone visible from Earth. They certainly may be traveling fast, but we cannot see them coming until they've crossed the singularity, and, as is the case with universal expansion, the light acceleration change allows *gravitational lensing* to always occur, with no method of cheating available for us to even briefly see into the futures light cone. Even if the rules of general relativity were different and light could momentarily travel faster than itself (and not just in cases of special relativity) our colleague couldn't see into our light cone existing in the future for even a second while they approached the singularity. A less scientific way to imagine this is picturing an underground subway station. It has a track with trains potentially moving in either direction on it, however, until the train physically arrives in the station, you cannot see it. The case of the eager traveler who walks past the threshold and looks down the dark subway tunnel to see the train coming just before it'll arrive in the station cannot ever occur in this illustration; the eager traveler would simply see nothing but black down the tunnel until the train arrived in the station, which depending on how they were positioned, might be hazardous to their health! Even if the train car in the front had an extremely powerful laser shining straight ahead (like a searchlight) down the path of the track in front of it, its beam would not penetrate or be visible until arriving in the station. The powerful laser wouldn't seem so mighty to observers in the station either, since it would only shine to the end of the track in the station and stop, seemingly midair.

In explaining why universal expansion is occurring, the preceding discussion about black holes was necessary; they're the principal cause both for the universe expanding, and, the ever-increasing expansion rate over time. Once matter has passed into a black hole and vanished from the light cone we can observe from on our side, it continues its increased velocity and more importantly begins existing in a different time. The faster information passes through a singularity, the faster it must then travel when "it's absorbed" by the virtual sponge of the black hole. The faster information travels towards, at and exceeding the speed of light, the faster it travels into time.

Most of the universe does not exist in the past after coming from the present or future; it's a mere anomaly that shares the same parallel with special relativity that general relativity does with nearly all the universe. It's imperative that special relativity exceptions are used to explain the phenomenon, as most physicists agree that it's impossible to defy *Newtonian* laws and have information (of any kind) simply disappear; they can change state in infinitely calculable ways but are never gone for good. The problem of light needing to physically travel faster than its general speed is a special problem, which only exists within a singularity and the result of the intensive gravitational intake of the black hole. The stronger the gravitation of a black hole, the quicker it accelerates lights incoming velocity, which is caught in the singularity and moved into a different light cone. Information would appear to disappear completely if you could observe a black hole sideways and from both light cones, which we know isn't physically possible. The information would accelerate and then simply disappear completely; the same way a coin on the floor is sucked into a vacuum cleaner when it's run over it on the floor, for example. This helpful illustration doesn't, of course, account for the fact that the information that has "disappeared" still exists, but in a different time and cannot be seen.

The next problem that needs explanation is what the information is doing that's been consumed by the black hole and now exists in a different time, relative to ours on Earth. It's not waiting with a cocktail for the rest of the universe to "catch up" to its very existence. The information is stretching the virtual fabric itself of the universe; because of the time dilation and intensive *gravitational lensing*, which begins at the light cone exchange point, which is generally the singularity of a black hole. Special relativity allows for bending of light and the *Schwarzschild radius* itself proves that a black hole is circular (in shape) from its intensive gravitational pull. [5]

Light bends as it is accumulated by a black hole and continues to do so at a linear rate, which follows its increase in speed, especially when it exceeds 186,000 miles-per-second. Instead of nice straight lines which can be visualized like flashlight or laser beams, they're rounded and twisted like metal in a car that's been flattened in a junkyard.

The faster this information goes, the more it bends and stretches as well as it's moving into the future from the present. Light bending and accelerating past its typical speed because of the massive gravitation of a singularity and moving into a different time is a messy affair; even though it appears to us that it's simply vanishing like water spiraling down a drain. Just as water from a drain ends up somewhere, so must the information that has become a twisted mess that does not even exist in the present anymore. Instead of flowing out to a holding tank or sewer, the information consumed by a black hole is stretching the fabric of space it occupies, which causes the universe to expand as the rate of consumption increases by the overall number of black holes which exist in the universe at any one time, plus the fact that as the universe expands faster and more matter exists, the chances for more new black holes increase, albeit always small, but ever steadily increasing over time. This progression represents the expansion of the universe and the dirty little secret black holes have been hiding for some time, no pun intended.

CHAPTER 3

HOW THE UNIVERSE IS EXPANDING

Universal expansion, as discussed in chapters one and two, is the phenomena which is causing the universe to expand. This occurs at an ever-increasing rate, which will never stop; because black holes always continue growing in solar mass from the matter they consume, and, as the universe ages, more black holes mathematically continue to form. Once a few proverbial black holes are let out into the wild, it slowly reproduces in number, causing more black holes to form, and so on. Eventually, these black holes consume enough mass they begin to stretch the "fabric" itself of space-time. The easiest way to imagine this is to remember all the empty "space" between the stars (and other objects which can be seen) that scientists have no explanation or theory about its purpose. This is what's so often described as dark matter; we're reasonably certain it's something, but what that something is, nobody knows. This has prompted several *quantum theory* explanations to be thrown at it; none of which can be even partially deemed correct outside of experiments conducted in a particle accelerator…none of which exist on the planet which are large enough (and which use enough energy) to succeed. Even the massive, multi-country accelerator known as *CERN* has not had any success in the least bit in this area. Anyway, this "empty" fabric which we cannot see becomes stretched; just like a ballon gets as it expands. It takes some fury to stretch everything around you, including that which cannot be seen, however, black holes are deadly and invisible by their very nature.

This "stretching" is an inescapable paradox, which will only continue accelerating until the universe itself ceases to exist; it can never stop once it has begun. It's hard to say with precision when this occurred, but it likely corresponds with the initial expansion of the universe itself, or shortly thereafter. While the explanatory detail provided in the first two chapters explains this phenomenon and the mechanics governing it, this chapter seeks to explain how to chart, measure and model how black holes are "stretching" space empirically; with individual black holes, groupings of them, or, all of them which are observable from Earth. This can be difficult and tedious to do even with a calculator, because both the distance and size of black holes can be very large numbers; which are unwieldy to work with, and, difficult for illustration purposes because of their variance and size. This is accomplished using a formula, which produces what's known as an *Eastman constant.*

The *Eastman constant* is a formula based on the distance of a black hole from the Earth, calculated against the size of its solar mass. The larger the resulting number is, the further the black hole has moved and subsequently, led to the universes further expansion, to a point. It's helpful for visualizing the expansion and growth movement in charting over time, and, provides smaller, more manageable numbers than the huge ones inherent in the distance and solar mass of a black hole. The purpose for developing a constant was to allow me to escape the calculability problem of having so many huge numbers, yet still allow for simple math to be done against them; allowing for charting and visual representation modeling to be done easier, and, without having long exponents attached to every number.

The value of using an *Eastman constant* for visual data interpretation is realized when comparing earlier (or later) values against the current observation. This shows the expansion byproduct of a black hole in the segment of the universe it occupies. When one expands the data set (to include mostly all known black holes that have been discovered from Earth, as listed in the Appendix section) against their respective solar mass, one may derive the overall constant of the universes expansion…because of the continual (and collective) consumption of matter. It's important to note that the *Eastman constant* is only reporting on the observations from "our side" of the light cone; the movement is dramatically less than on the other side, and also, because matter that's velocity doesn't exist faster than light will slow.

Since not even light itself can escape the pull of the event horizon in a black hole, this ensures a constant influx of solar mass, which can be measured across time using radio telescopes. It's easier to understand this constant by imagining a huge vacuum cleaner, which represents a black hole that's been switched to its maximum power setting and is never, ever turned off; causing even matter which falls (or travels) near its reach to also be eventually consumed. Because the universe expands much faster because of information consumed at more than 186,000 miles-per-second on the inside of black holes' collective light cones, the rate calculated over time using the *Eastman constant* is necessary always a *Newtonian mechanics* exercise, it can never be quantum, but still reflects the much smaller linear gains that are feeding the universes quantum acceleration; due to the infinite properties of matter traveling faster than the speed of light. Effectively, a much smaller *Newtonian* measure is occurring for one segment of the observable universe that's responsible for the same exponentially quantum expansion event.

The *Eastman constant* is calculated using the following formula:

$$Dc \times V\left(\frac{M}{\pi}\right)$$

Dc represents the distance of a black hole from Earth in light years.

V represents the speed of light, which is 186,000 miles a second.

M represents the observed solar mass of a black hole.

As an example, let's consider the black hole located in M32. It has a distance of 2,400,000 light years from Earth and a solar mass of 3,000,000.

$$2{,}400{,}000 \times 186{,}000 \times 3{,}000{,}000 / 3.14$$

Once the arithmetic is completed, the values appear as:

$$446400000000 \times 955414.01273885350318$$

$$= 4.26$$

In the case of the black hole M32, the *Eastman constant* is 4.26 and it may be observed over time to calculate the *Newtonian* expansion caused by the black hole. If the *Eastman constant* is next measured (for example) at 5.12, then the differential of 0.86 represents the observable expansion that's occurred. Having an empirical basis for measurement over time helps us calculate the expansion differential caused by black holes from our position on Earth with more confidence.

Since all black holes are circular (even if not immediately visible to us) it's necessary to use the value of pi for division of the solar mass; as Earth is also round and this calculation measures the distance between two bodies which are always round. Since black holes are quite distant from the Earth and can often be rather large, the use of many zeros is always a real factor, which the *Eastman constant* helps to alleviate in the use of more manageable numbers.

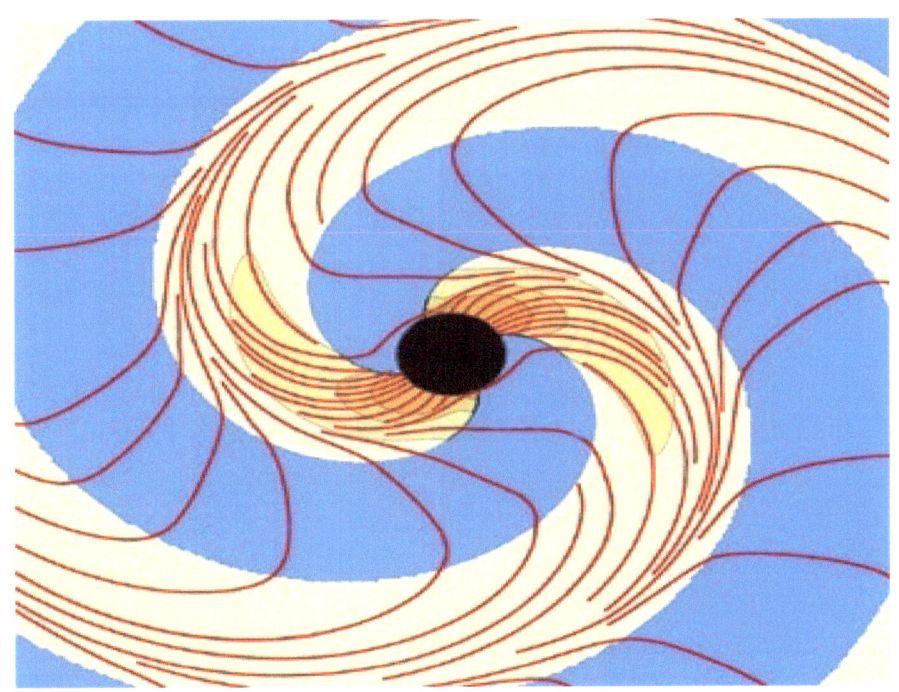

Vortex lines display the spiraling gravitation effect of a black hole [6]

In the case of the overall landscape of all black holes which are observable from Earth (which is approximately 90) we can calculate the *Eastman constant*, which shows the universes overall expansion over time; at least from what we can see of it with radio telescopes.

27,741,714,750 x 186,000 x 72,436,356,348 / 3.14

5159958943500000 x 23068903295.54140127388535

= 3.69

It's imperative to always use the average of two solar mass figures, if they have a published variance for accuracy over time. While the distance from Earth is typically known more precisely, the solar mass of a black hole can sometimes be difficult to measure. For example, if a black hole has a solar mass of 8,000,000 to 10,000,000 than it should be calculated as 9,000,000 for each one containing variance. It's also important to use the same calculator or computer for doing these calculations, as this helps eliminate precision bias and allow for more accuracy.

Next, we can experiment and determine what the overall *Eastman constant* is for all observable black holes from Earth at a later point in time:

$$741,714,950 \times 186,000 \times 102,436,356,348 / 3.14$$

$$137958980700000 \times 32623043422.92993630573248$$

$$= 4.50$$

The differential between the time of this book's publication and this example of a future *Eastman constant* calculation is 0.81 and represents the difference in expansion of the universe from our perspective on Earth only; we don't have data for every black hole which exists and can only use everything we can observe from Earth. Obviously, the larger the differential from the last calculation, the more expansion is occurring, whereas a smaller number indicates less expansion.

It's instructive to note that the calculation of an individual black hole may have a higher *Eastman constant* than a weighted average of everything observable from Earth. The "now and later" comparison can be performed just as easily on individual black holes as well. Now, let's return to the individual black hole M32 alone, but add some additional solar mass to simulate passed time.

$$2,400,000 \times 186,000 \times 4,405,000 / 3.14$$

$$446400000000 \times 1402866.24203821656051$$

$$= 6.26$$

This approach allows for charting at any time interval (or

series) while keeping the data relatively manageable in visual depictions. The distance from Earth is subject to change in the same manner solar mass does over time, however, it's typically not anywhere near the change agent (over time) that solar mass is. Black holes are no doubt responsible for the expansion of the universe; however, their solar mass moving through the event horizon is what's increasing over time.

It's reasonably simple to visualize that the solar mass of black holes (irrespective of their size) far eclipses their distance from Earth in light years. While still being much too far away for us to visit with a satellite, for example, they're still proportionally far larger than they are far away from Earth. This point helps to underscore both the incredible power and size of black holes, and, the "stretching" from their increasing mass being continually expanded; continuing the acceleration of the expansion of the universe for perpetuity.

The table on pages 33 and 34 illustrate all supermassive black holes which can be detected from Earth as of publication. The growth they'll achieve (at their current rates) in one year and then 100 million years is shown on the right. This is a sobering reminder that all black holes, irrespective of size, will continue to accumulate solar mass for a practically infinite period. This causes the "fabric" of the universe to metaphorically continue "stretching" and thus representing not just the expansion of the universe, but also it's ever increasing rate of expansion, which will never stop until the universe simply ceases to exist.

The *Eastman constant* provides a simple means to chart and model the progressing expansion of the universe; either at the individual black hole level, or, other groupings of interest. Obviously, the larger the black hole, the quicker it will expand and further grow, effectively "eating" more of its neighboring matter than a much smaller black hole which gravitational effects aren't as pronounced. One way to visualize this effect is to envision a huge balloon inflating with all the solar masses of all black holes in the universe (including all the ones that we cannot detect) as they're "sucked in" to the singularity of each in real time. Obviously, the balloon will continue filling faster and stretching its fabric to hold as much matter as possible; the expanding balloon is representative of our ever-expanding universe. The balloon in this example is special, however, and doesn't explode when overinflated; it simply continues to expand continually, with new fabric being added as needed…the balloon continues to inflate forever. The balloon ceasing to inflate, in this example, would be equivalent to the end of the universe.

	Solar Mass	Growth Per Year	100 Million Years
3C273	1,000,000,000	1	1,100,000,000
ARP 151	13,000,000	1.3	143,000,000
IC 1459	1,000,000,000	1.4	1,140,000,000
M104	660,000,000	0.8	740,000,000
M105	130,000,000	0.4	170,000,000
M106	31,000,000	1.7	191,000,000
M31	30,000,000	1.2	150,000,000
M32	3,000,000	1.5	153,000,000
M51	1,000,000	0.9	91,000,000
M60	4,500,000,000	1.3	4,630,000,000
M77	15,000,000	0.8	110,000,000
M81	7,000,000	0.4	47,000,000
M84	1,500,000,000	0.7	1,580,000,000
M87	6,600,000,000	1.3	6,730,000,000
MILKY WAY	4,100,00	0.7	84,100,000
NGC 1023	50,000,000	1.1	160,000,000
NGC 1194	65,000,000	1	165,000,000
NGC 1277	17,000,000,000	1.6	17,160,000,000
NGC 1365	2,000,000	0.5	52,000,000
NGC 2273	7,500,000	0.6	67,500,000
NGC 2778	20,000,000	1.1	130,000,000
NGC 2787	40,000,000	1.7	210,000,000
NGC 2960	11,500,000	1.2	131,500,000
NGC 3115	900,000,000	0.9	990,000,000

	Solar Mass	Growth Per Year	100 Million Years
NGC 3516	23,000,000	1.6	183,000,000
NGC 3585	340,000,000	1.4	480,000,000
NGC 3607	125,000,000	1.1	235,000,000
NGC 3608	210,000,000	0.8	290,000,000
NGC 3783	9,000,000	0.5	59,000,000
NGC 3842	9,700,000,000	1	9,800,000,000
NGC 3945	9,000,000	0.6	69,000,000
NGC 3998	570,000,000	1.2	690,000,000
NGC 4026	210,000,000	1.3	340,000,000
NGC 4061	4,500,000,000	1.7	4,670,000,000
NGC 4151	45,000,000	0.8	125,000,000
NGC 4178	200,000	0.5	50,200,000
NGC 4253	20,000,000	1	120,000,000
NGC 4261	800,000,000	0.9	890,000,000

CHAPTER 4

TIME DILATION AND UNIVERSAL EXPANSION

Time dilation is a complicated subject; one that an entire book or academic dissertation could easily be alone dedicated. It's primary importance to *universal expansion theory*, however, is serving as the primary reason why we cannot observe the actual "stretching" of the universe occurring from all the matter being consumed within a black hole's singularity. An easier way to envision this concept is simply that this "stretching" phenomenon that's accelerating the expansion of the universe is occurring beyond a glass door that we cannot see through. We can see that the door is made of glass from the outside, but we cannot see through it, not unlike a heavily tinted automotive window. Time dilation is the principal cause of the "vanishing" glass door that opens into a black hole's singularity.

If we could open the door and see inside, there'd be a massive influx of matter being compressed and moving beyond it at more than the speed of light, with most of this "total information" being consumed and spread into the fabric of the galaxy. We couldn't physically view any of the movement of this information / matter moving past the inside of the door because it's traveling faster than the speed of light, which is a requirement of universal expansion (as highlighted in chapter one) and is why only the color black is observed from a black hole, among other reasons already well understood. Time begins slowly moving backwards when approaching the speed of light and drastically slows (enough to the viewer in our dimension) when exceeding it that it simply cannot be seen and appears black. It's agreeably frustrating to "know" what lies beyond the singularity of a black hole and to have an empirical basis to prove its characteristics without being able to physically view it...the result of time dilations effects when exceeding the speed of light in a vacuum.

This sounds paradoxical at first, but it's exactly what's occurring from the eyes perspective, because we're always looking into a black hole's event horizon and can never see beyond the singularity. Even having a glass front door, or no front door at all matters not, as we cannot observe what we cannot see and the back door is not situated in the same dimension. We see everything from Earth (even with radio telescopes which are listening to faraway objects, to then build their images for us) in the third dimension. The third dimension allows for observing the front door of a black hole's house, but not it's back door...where the matter is being dispersed and is "stretching" both time and space, as well as driving the expansion of the universe.

As highlighted in chapter one, the number of black holes will only continue to increase as the universe ages and expands, and, each black hole (regardless of solar mass) is contributing to the "stretch" that may well have begun the overall universes expansion. Even a casual observer not skilled in astrophysics can deduce that as the number of black holes in the universe naturally continue to increase, they continue to accumulate and disperse matter beyond the speed of light over time, which further expands the universe, and finally, that this process continues occurring exponentially and will never end until the universe ceases to exist. This means the universe will not only continue expanding, but it will continue to do so at an increasingly faster rate. Imagine a giant spider web, which initially starts in the middle and continues its construction in all directions and never stops; with the number of spiders working on the web steadily increasing forever over time.

This can be espoused by the equation on the next page, which explains the reason we cannot observe through the "glass door" of a singularity in a black hole and "see" the information stretching the universe at above the speed of light; until it could decay somehow to at least the speed of light. Time dilation is working its proverbial magic with the increased speed of matter past light, as it's required for all light to escape a singularity as explained in chapter one. We cannot see backwards in time from our third dimension on Earth.

$$T = \frac{C^1}{M}$$

T represents time, which goes from slightly slowing from matter traveling near (or at) the speed of light as it enters the event horizon; to backwards from light exceeding its own speed when being consumed by the singularity.

C represents the speed of light in a vacuum increasing beyond its own speed.

M represents the quantum matter being consumed by the singularity and traveling beyond the speed of light, exhibiting an extreme time dilation shift.

The *information paradox* holds that quantum information is not destroyed or lost after it's disappeared into the singularity of a black hole, despite no longer being visible, and, having no proof that the quantum information still lives on. Several prominent physicists have attempted to solve this problem, with the loudest theory coming from Stephen Hawking, who famously declared in February 1974[7] that the quantum state of a black hole allowed for the destruction of information, which is known as *Hawking radiation*. Hawking missed *Liouville's theorem of mechanics*, which proves the distribution function is constant along any given trajectory in phase space and makes his theory extremely difficult to accept, given existing evidence.[8]

Gerard 't Hooft and Leonard Susskind famously disagreed with Hawking (as did I not so famously) and in 2008

Susskind published a book which gives a string theory basis to the *holographic principal*; which states that the maximum entropy in any region of an event horizon is squared and not cubed as in existing formulas.[9] This assumes the universe is a two–dimensional information representation and that the three dimensions we observe from Earth (for instance) exist only at low energies. This solves the *information paradox* and dispels Hawking's assertions about information being destroyed, despite a 2004 attempt to admit that information might escape from an event horizon due to *thermal conformal field theory*. Hawking moved further from the actual truth in accepting the overall assertion that it's doubtful information is lost within the singularity of a black hole, however, in disproving him, Susskind disregarded the *second law of thermodynamics*, as hot gas with entropy would certainly disappear once it crossed the event horizon. Jacob Bekenstein demonstrated in 1981, for instance, that black holes have more entropy than anything with a similar volume, and, that it's directly proportional to its event horizon.[10]

Another bit of importance that can be attached to the idea of a *holographic principal* is for perceiving instead that the universe is nothing more than a series of virtualized computer applications; running somewhere on massive computational engines, which are under intelligent control…or at least were at some point during their inception. An easy way to visualize this notion is by watching the hugely popular Matrix film trilogy, which features an architect and several human–looking machines which administer and power an entire world; by using humans solely for their bodies energy, leaing them in a state of suspended animation. This causes them to experience life without having the notion that they're existing solely in their own imagination, and, that everything in the entire galaxy is physically real…when it's simply been virtualized. This concept has been gaining popularity in recent years and has been no stranger in the quantum physics community either. While most agree we simply don't have the capacity yet to determine if our universe is physically real or just virtualized to appear that way, the great news is that it doesn't matter for universal expansion! Regardless of which style one prefers for explaining the composition of the universe itself, the *theory of universal expansion* holds true in either case. So, regardless of which eventuality may be the correct one, it matters not in the case of why the universe is expanding, and, doing so faster.

One way to accelerate research and make progress on big ideas is to collaborate with others who might hold even a shred of belief (or interest) in proving or denying your work. Being outside the realm of academic physics, this approach seemed correct and worth pursuing…until finding that not many folks had any interest in *universal expansion theory*; not to mention actually contributing to it. The list narrowed to quantum physicists who had challenged Stephen Hawking in the past. I curated this list further to include just black holes and wasn't surprised to find only one name left to consider. It was somebody who's work I enjoyed tremendously, and, lived close to me. This seemingly positive irony made me quite excited…like a child about to open a gift. It was time to do a fresh review of all this most interesting persons published work and then reach out to them. I'd always wanted to talk to this person and became convinced they were the one person who'd be willing to get their hands dirty with my theory, and, even better, they'd enjoy doing so. They would be fearless with the "paradox problem" and wouldn't think I was crazy.

I then tried my luck at this ongoing "paradox problem" with Leonard Susskind in a physics class at *Stanford University* in 2013, realizing the topic being discussed at the moment was somewhat related to my *universal expansion theory*; which holds that information consumed by a black hole's singularity is not lost and is instead pushed into the vacuum of space at above the speed of light, which causes time dilation and the inability to observe the "stretching" of the universe as a result. Dr. Susskind disagreed instantly with my theory that black holes might be responsible for the increasing expansion of the universe. He almost seemed annoyed with my question, which I'd patiently waited several weeks for the class topic to finally relate to directly before asking. "No, we've looked at that before and it's not relevant," was the thrust of his response. I was sad that the one physicist who I expected to take my theory and run with it was very much not interested. After all, this was the person who won a public debate about black holes with Stephen Hawking, and, both are extremely brilliant. The depressing drive home from Palo Alto that night fueled me to further refine my *theory of universal expansion*; so, that others might be able to make their own determination, based on the findings and research.

CHAPTER 5

RELATIVITY AND UNIVERSAL EXPANSION

Einstein famously demonstrated some hope for time travel with his equations for *general relativity theory*, but the larger breakthrough in this area came in 1916, when Ludwig Flamm discovered what's known as an *Einstein-Rosen Bridge*, or in more contemporary terms, a wormhole that allows travel in one direction. [11] Since then, science fiction films, novels and radio dramas have all shown us the majestic idea of using such phenomenon to serve as a personal time machine; with the ability to go backwards and forwards in time merely by flying your spaceship towards the left or right of a giant glowing vortex. Unfortunately, this wasn't exactly what the physics community was necessarily touting in response to the revelations gained from Einstein's *special theory of relativity*, arguably the most famous of his incredible and plentiful career.

Special relativity was first published by Einstein in 1905 and has two principal rules; that the rules of physics are identical in all states of common motion in a time-independent manner, and, that the speed of light in a vacuum is the same for any observer, even if the lights motion is varied. [12] The prediction of the equivalence of mass and energy is perhaps the most famous equation of all time (no pun intended) and the product of the two rules of special relativity. Special relativity allows for different rules as the lone exception in astrophysics than the conventional "Standard Model" we use for predicting and solving physics problems. In other words, special relativity presents the rules for exception cases in the galaxy, with the largest suspect being black holes. It also explains why some phenomenon in the cosmos doesn't appear to be following the same rules that everything else consistently does.

The most important revelation that most people learn in school about Einstein's work is concerned with general relativity, which he published (seemingly backwards) in 1915 after his work on special relativity; this is that light can bend due to gravity, which is known in physics as *gravitational lensing*. [13] This happens when light is absorbed by the gravitational pull of a black hole, but can happen also by viewing light from distant galaxies; the gravity of matter between our planet and the distant galaxy causes light to bend. This causes what appear to be multiple images of the same thing. Distant photos taken with the Hubble telescope often show this characteristic.

Another important principal from general relativity that some learn next is what that famous formula Einstein's most known for is actually concerned with. This is the principal that allows a nuclear explosion to occur and is also why when the velocity of any object increases, its kinetic energy also increases, which causes its mass to increase. Einstein was proving that energy and mass share the same relationship that excitement and fear do; they're simply derived from the same place, but have different outcomes.

Finally, the basis of general relativity and the rule of special relativity itself stipulates that the rules are the same for all observers, and, that the speed of light was constant and didn't change as it was traveling (as previously thought) through the universe; the motion of what started the light on its journey and the casual observer on Earth, for instance, didn't cause fluctuations of speed. Thus, everything in the universe is moving relative to everything else; the principal that relativity was named after. The work of special relativity, while published earlier, offers an even more extensive basis for dealing with black holes.

Perhaps the most important element of Einstein's work with respect to universal expansion is the exception case of a particle traveling faster than the speed of light. Intensive gravitation from a black hole causes even light to accelerate faster than its regular constant as part of the basic "greater or lesser than" property that the singularity of a black hole exhibits. The often-exponential acceleration that occurs causes the resulting space-time fabric to "stretch" as the speeding information exits the black hole. Special relativity provided the first basis for an exception case where matter could travel faster than the speed of light. The principal of time dilation was also an important revelation of special relativity; this held that all of space is like a grid and that time is necessarily relevant to the observer. This means that something that occurs for us at one time could occur at a very different time for somebody else.

Time effectively was established as a fourth dimension that is always occurring relevant to the individual conditions of the observer. If you were traveling at the speed of light, or even remotely close to achieving it, your observations of the same phenomenon would be different than somebody traveling much slower, or not moving at all. Time passing slower for those traveling faster, and, the popular example of two wristwatches set to the exact same time having different times displayed on them when one is given to a person traveling near the speed of light is this principle in action. Thus, Einstein also developed the principal of time dilation, which occurs when the velocity of a particle approaches or exceeds the speed of light; per universal expansion, this occurs when information is consumed by a black hole. This is because the information typically increases speed beyond the speed of light to be consumed by the ever-growing gravitation of a black hole. It also serves as the basis for explaining why we cannot ever observe what occurs "on the inside" of a black hole, or even on its imaginary sides; like corners of a box extending around the singularity. Humans on Earth are observers forever trapped on the outside of the black hole and as such, we cannot see into the future that's occurring inside of it, or on the "other side" that we might refer to as the back door. Consequently, if we were observers who had traveled into a black hole (we're ignoring the fact that we couldn't physically survive the encounter for the sake of convenience) we wouldn't see the same things as those who're still on the outside (quite literally) looking inside the black hole.

Special relativity generally states that mass cannot travel faster than the speed of light because the mass and momentum would grow infinitely, however, it is possible within the constructs of special relativity to have mass who's velocity exceeds the speed of light; if it necessarily maintains that speed forever. A particle traveling faster than the speed of light is known as a tachyon. Gerald Feinberg coined the term tachyon in 1967, as part of his work in *quantum field theory.* [14] The information that's consumed by a black hole becomes effectively consistent with a tachyon and never experiences any information loss; it only continues infinitely, with the laws of motion being satisfied because the consumed matter never loses its original energy and achieves truly perpetual motion. Matter which then becomes tachyons not only passes each test of motion with flying colors, it even earns extra credit; because the starting matter in this case could then only accelerate faster than the speed of light before reaching this perpetually infinite state. Motion is conserved because it's never effectively lost.

This explains the inability to detect or view tachyon information by any optical or radio means from Earth; the matter involved itself is subject to time dilation from its extreme velocity. As such, it'd be completely invisible to us as it approached, even if we were physically located next to its travel path. From a special relativity standpoint, the information contained in a tachyon would be necessarily existing in an earlier time than is occurring on Earth, but never in the future. The mass itself is traveling faster than any light contained inside it; therefore, we could not see it coming until it had already passed us, not unlike the principal which occurs with the *Doppler effect* when a fast-moving jet aircraft approaches and can be seen a few seconds before it's heard. Per the principal of *Cherenkov radiation* (which was experimentally detected by Pavel Cherenkov) once a tachyon had physically passed us, we'd see two different images of it; one would be approaching us and the other would be simultaneously departing. [15]

When a tachyon could fundamentally loose energy because of *Cherenkov radiation* when exceeding the speed of light, it conversely increases speed when its energy is diminished. Physics loves to use copies and pairs of information when the going gets weird; which is no different in this situation. The singular matter that exceeds the speed of light develops an oppositely charged copy of itself, which always "balances" them both without either retaining any energy in the infinite speed that ensues. Thus, the momentum of the (now two) tachyons can be infinite in either opposing direction; this eliminates the apparent paradox of the law of conservation. At the point which both tachyons reach infinite speed at the exact same point in space-time, they cancel each other's effect and the "copy" is then reduced back to a singular tachyon. Matter which contains gravitational mass always experiences an increase in speed as it travels through space-time; even if such mass encounters different kinds of particles, it may pass on *Cherenkov radiation* to those particles and then continue its journey.

Calculating how a tachyon behaves was one purpose of the *Hamilton-Jacobi equation* [16] in physics. Tachyons are particles that have "imaginary" mass known as rest mass; this property of special relativity exists to account for the problem of mass achieving a velocity exceeding the speed of light, as discussed in chapter five. The velocity of a tachyon is thus always going to exceed the speed of light when it emerges from a black hole. The rest mass of a tachyon is expressed [17] as:

$$m_0 = im.$$

The value of m★ is always a positive number. Tachyons' energy have a velocity that exceeds the circumference and is generally expressed in a formula as:

$$E = \frac{m * c^2}{\sqrt{v^2/c^2 - 1}}$$

A tachyons energy decreases as its velocity increases and for the sake of formulas and mathematical calculations, its velocity value becomes zero as a limit of infinite velocity, as explained by Vladimir Lipunov in his 1978 paper on tachyon motion, cited above. At this point, a zero-energy tachyon is transcendent and will continue in this fashion forever, since tachyons velocity cannot ever slow down and are infinitely stuck at a speed above the speed of light.

Special relativity was published by Einstein in 1905, immense work was done by Kerr, Schwarzschild and many others since, and, the equations necessary for illustrating that tachyons cannot slow down (and infinitely continue above the speed of light) above were published in 1978. How nobody could have concluded black holes are inherently responsible for the expansion of the universe, and, that the information consumed by their singularities becomes converted to tachyon particles (which never stop traveling faster than the speed of light) is baffling; especially with so many successful redshift measurements being performed in the last century that illustrate that the universe is expanding at an ever-increasing rate. Until the quantum physics community began addressing the problem Hawking incorrectly theorized about with information being destroyed in black holes, academia was somewhat going backwards on itself; challenging the established body of work from the last century in strange ways that read well for a graduate college thesis, but did nothing to advance the study of black holes and continued moving them backwards instead; with the evidence sitting neatly and prominently available, which demonstrates clearly how information converts from *Newtonian* motion laws to quantum ones when consumed by a singularity, and, that it becomes a tachyon…which never stops traveling faster than the speed of light. It's hoped this work will help put a more attractive look at these past shortcomings and advance the study of black holes and universal expansion forward, instead of backwards.

Tachyons were thought to have been detected for the very first time with a tau neutrino experiment in the particle accelerator at *CERN* in 2011, however, subsequent reexamination of the evidence later proved to be, "a faulty element of the experiments fiber optic timing system." [18] Nonetheless, the search for empirical proof of tachyons continues; some in the theoretical physics community have expressed doubt about their existence, however, we know from the principles of relativity and time dilation that some phenomenon is difficult, if not impossible to detect or view from Earth. The long hunt for the elusive *Higgs boson* particle immediately comes to mind when considering tachyons and their role in universal expansion. While one may not immediately be able to view them, it doesn't mean they don't necessarily exist.

CHAPTER 6

HAWKING RADIATION AND UNIVERSAL EXPANSION

Dr. Hawking's brilliant discoveries in physics and with black holes lead to the principal of *Hawking radiation* in 1974. This suggested that information has a virtual "photocopy" made of it just before it enters the event horizon of a black hole. The resulting copy escapes and is spared from the intense gravitational pull of the black hole, while the original information is consumed. Hawking further explained that over time, a black hole may simply "burn itself out" like a candle reaching the very bottom and cease to exist, with information being emitted from it like heat dissipating from a hot roof in the summer. The idea that every piece of matter having a twin created which then escapes the singularity of a black hole when its consumed was controversial; suggesting more importantly, that information could escape the gravitational pull of a black hole, even if it was traveling at the speed of light. If that wasn't enough controversy, the idea that black holes could simply evaporate over time and dissipate information of any kind added even more to the physics community.

The discoveries and impressive body of work Stephen Hawking's made in physics have certainly accelerated both the interest and understanding of black holes for academic and non-academic audiences alike and should be commended. Most people outside the field of physics were not as interested in the strange phenomenon of black holes until his work emerged and was published, which has been most fortunate; even influencing others to pursue physics when they may not otherwise have done so. The notion that a copy of information was made prior to entering the event horizon of a black hole has intrigued me for many years, becoming an interesting thought exercise for me in theoretical physics. It's easy to start theorizing where the phantom copy is going and where it ends up, or, why it happens to begin with. Unfortunately, the notion that copies are made of all information that's consumed by the singularity of a black hole may be both inherently impossible and false, as this chapter shall now explain. The theory that black holes may dissipate and effectively cease to exist is certainly most interesting and may have merit, however, it'll be difficult to observe empirically, and, whether they cease to exist makes no difference with respect to the outcome of universal expansion.

One of the biggest problems with *Hawking radiation* is that it's erroneously thought to be a veritable "burp" of matter emitting from the singularity of a black hole. This fascinating notion has caused fervor in physics because of its other assumptions, as well as the problem itself of matter exiting a black hole through the same door that is used to enter. The problem of having two copies of all information (the photocopy effect) is perhaps the largest "sub-dissent" amongst *Hawking radiation* critics and yet it was still a brilliant idea to explain the problems inherent in having information escape from the front door of a black hole.

Of course, if one makes the logical conclusion that *universal expansion theory* is correct, it's far easier to reveal the existence of the completed story and miscues inherent with *Hawking radiation*. First, it's necessary to note that the phenomenon Hawking describes is occurring, however, the information is exiting the black hole from the back door and not the front door; where the singularity is. The second item of importance is that no information photocopier exists in the event horizon (or even anywhere near it, for that matter) that's actively copying matter before it enters a black hole. It matters not if the information was entering the black hole using the front door, back door or even a half-open kitchen window; it simply isn't being replicated and that's because (at a simplistic level) it's not necessary and at a higher level, because it cannot.

The magic of using the back door of a black holes proverbial "house" is camouflaged to us because matter approaching and exceeding the speed of light suffers time dilation and begins existing in a different time, whereas as observers from Earth, we're stuck in the present and cannot see into the future; we have no crystal ball in astrophysics. The radiation Hawking envisioned being emitted from the singularity of a black hole and dissipating (like heat waves radiating from a kitchen stove) is occurring in a different manner than he realized; he was on the cusp of solving the bigger problem without accounting for time dilation and where the information itself was going. The "why" element simply wasn't explained by the information photocopier theory, which whilst convenient, is false. Hawking was merely trying to solve the problems of black holes using only the front door we can observe, and, with the assumption that the front door could be used to exit as well as to enter the black hole, irrespective of copying information; this was coupled with the belief that light couldn't exceed its own speed in a vacuum, and finally, the role time dilation plays in obscuring completely the front door and all the information "guests" who step in it from outside. We can see our information guests stepping inside the black hole using the front door located in the singularity; however, we can never ever see them leave, even if we wait longer than both the black hole and our house guest could continue to be alive or exist. We must presume that either no guest ever leaves the black hole house, or, that they may only exit through using the back door, or, a window that's someplace on the backside of the house. Hawking correctly assumed that information guests were, in-fact sometimes exiting the house, however, his error was believing that they used the front door to exit, and, in trying to explain this difficult paradox, that copies of the guests were magically made after they wiped their feet on the doormat and stepped inside the threshold of the front door. This is a very difficult problem that's not at all obvious, and, since the discovery of black holes, the little we've detected of them from Earth has shown that they seemingly have very attractive looking fronts to their houses; lots of curb appeal and no neighbors, but, we never, ever see the backyard. Buying a house without seeing the backyard, or, ever being able to ever leave again through the front door is an experience most of us simply wouldn't want anything to do with. After all, who's to say when we enter our new "house" that we'll ever be able to exit it again, and, if we do, we might not be anywhere near the city or country it was located it. That makes the work commute and child rearing very difficult, if not impossible! Most of us would simply cross that area off our maps and look for housing elsewhere; in a place where time didn't change, space-time rules weren't bent and we could physically use both our front and back doors, and, see the backyard from the street.

A phenomenon known as extragalactic jets is thought to be energized plasma in a focused beam. In the case of NGC 3862 (which is the sixth brightest galaxy from Earth) there are observable extragalactic jets, which were discovered by the Hubble Space Telescope in 1992. The black hole thought to be located somewhere in NGC 3862 was captured (below) in video frames throughout 20 years of archived Hubble footage by Eileen Meyer, of the Space Telescope Science Institute in Baltimore, Maryland with surprising results. The observable extragalactic jets appear to both catch-up and collide with each other, with some of them moving at 7 times the speed of light. 1,302,000 miles-per-second (or 2,095,365 kilometers-per-second) is quite fast indeed! It's faster than light is supposed to travel, however, it's commensurate with Einstein's rules of special relativity, which allow light to exceed 186,000 miles-per-second in special circumstances. One such circumstance appears optically for the first time to be extragalactic jets emanating from a black hole over time. Video on NASA and the ESA's websites show a 20-year time lapse of these jets via images taken with the Hubble space telescope in

NGC 3862. The electronic version of this book shows this short video, which appears as a general "pulsing" that moves slowly from where it appears to begin. It takes 20 years of observation to notice the subtle change occuring in one point in the sky that's thought to contain a black hole.

The following chart displays these jet movements, starting in 1994 and ending in 2014, as well as a "zoomed out" photo of the region (on the left) in the galaxy where NGC 3862 resides, which lies in the constellation of Leo.

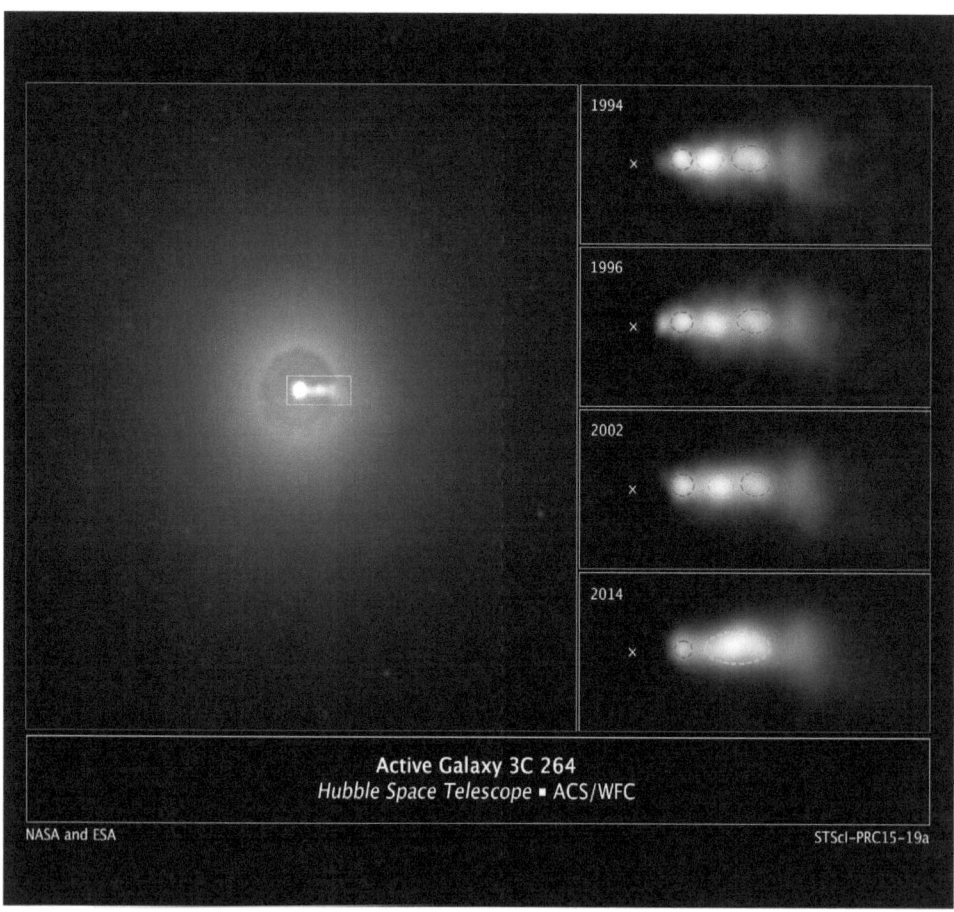

Courtesy: NASA, ESA and E. Meyer STScI

Visually, the real-time expansion of the universe seems to first resemble a rocket thrust tail as it shoots upwards through the atmosphere. Upon closer inspection, however, the white "pulsing" jet streams resemble the intake and resulting outtake flow of blood from the hearts pulmonic valve; the time lapses show an almost heartbeat-like quality one might expect to see on an echocardiogram or magnetic resonance imaging of a human heart. The universe has a veritable heartbeat while its expansion is occurring. The information consumed by black holes eventually accelerates to faster than the speed of light, and, because of developing into a tachyon and exhibiting *Cherenkov radiation*, it will never stop and simply exist as part of the fabric of space in the present time wherever it has directly passed as it continues. The means to accelerate even faster necessarily exists, however, as discussed in the last chapter, once this great of a velocity has been achieved in the vacuum of space, it cannot ever slow down or stop. Light only slows its velocity in the vacuum of space when it passes through an object, generating a refractive index that can be used to calculate its slowed velocity. While the expanding information is traveling above the speed of light as a condition of both special relativity and being consumed by the singularity of a black hole, it is also colliding with itself, as not all the information (or light) was traveling the same speed when it was "sucked in" to the singularity and then increased by the amount of massive gravitation inside the black hole. As black holes increase in solar mass, their gravitation or "suction" increases and the speed at which information accelerates as it's captured by the singularity also increases. It's important to note that it may take a very long time for an extragalactic information jet to be more easily detectable from our space-time on Earth.

Only some of the information that's participating in this expansion process is indirectly viewable from Earth, even using the Hubble space telescope over a 20-year period. The extragalactic jets which are viewable in the Hubble telescope animation (available in the electronic version of this book) are white in color and indicative of light experiencing absorption from other similarly captured light. They can be observed as white in color because all the wavelengths are still present after being absorbed by other fast moving streams of light. When light absorption occurs, the wavelength of the light also decreases as its velocity decreases. The accepted standard formula (below) for calculating refractive indexes explains this process empirically, with C representing the current speed of light, and C_m representing the speed of that light in the material through which it passes. C_m can never be a value that's higher than the speed of the light in C in this formula using either *Newtonian mechanics* or general relativity, so, the equation always has a positive value of 1, or more. As the light travels through objects with higher density, its velocity slows. This occurs both with light collisions and with other matter, such as gas, meteors, rocks, etc. In reality, once information has been accumulated by a singularity, a negative result would always occur, due to the velocity exceeding the speed of light.

$$n = \frac{C}{C_m}$$

Time dilation's also inherently to blame for some of the disparities in viewing captured information from a singularity expanding into the universe at a high rate of speed, often exceeding its own regular maximum in a vacuum. It's also why the eventuality (and now logical progression) of black holes being associated with the expansion of the universe has taken so long to be discovered. The rate at which the extragalactic jets of emanating information captured from a black hole accelerates, both as they enter the singularity and when exiting via the "back door" of a black hole is dependent upon more than one factor. The complication of not being able to see the matter exiting the black hole (from our position on Earth) makes any perfect set of calculus in this area difficult, however, the laws of motion, special relativity and the use of statistical mechanics are helpful for discerning a fair approximation of the phenomenon of matter exiting a black hole and stretching the fabric of space simultaneously in different times to the same observer. Light absorption is ultimately one of the key events which slow the speed of light in a vacuum, unless it has, of course, eclipsed the speed of light and no longer applies. The universe, in this case, will be expanding far, far more rapidly in 10,000 years than it is now, because there's be simply nothing to stop it; in the same way filling a balloon with a helium tank will always produce an exploded balloon if the helium flow isn't stopped at an appropriate time during filing. In this example, the expanding universe as a helium tank will always be continually filling a giant balloon that can never fully inflate, and, the helium tank itself never becomes depleted.

Light itself is simply electromagnetic radiation which exists in (usually) several wavelengths. A different sound in different waveforms on an analyzer device (such as an oscilloscope) is like the principle that different color spectrums exhibit with light wavelengths. The excellent part about radiation is that we've become quite adept at measuring it, both here on Earth with Geiger counters, and, in distant space. The not so excellent part is that our own atmosphere hides most electromagnetic radiation from space; except for the light in the spectrum that's visible, and, some infrared light. Radio telescopes have helped to overcome our atmospheric "blocked view" that's no different than a very tall person sitting in front of you at a cinema, or sporting event. You'll certainly be able to see some of the motion (or action) occurring on the field, but not all of it, and, not all of the same spectrum that you'd have if no person was seated in front of you. Optical observation using the Hubble telescope, for instance, rectifies this problem and allows for viewing of radiation, whereas the largest percentage of observations are generally done with radio telescopes and then visualized (using a computer) into an image that the eye can discern. This has helped considerably in both the discovery and measurement of black holes, usually by observing strange changes occurring with their celestial neighbors. One other method of measurement is the intensity itself of the radiation, which is typically done by assigning colors to established common temperature ranges and then combining the data into a single image. Observing phenomenon related to black holes and supernovas thus becomes a detailed listening exercise, with the results printed out for our visual interpretation by colored pixels on an image. Rarely does the 20-year optical time-lapse photography exist with black holes as with the Hubble telescope example earlier in the chapter! Most measurement and validation of theories for black holes and supernovas are done by listening and visualizing the radiation; with heat being the most dynamic change agent, particularly over time.

Now that radiation and heat has been properly explained, thermal fluctuation is important to consider next, with respect to the movement and speed of extragalactic jets. Heat is the product of collisions and generally friction for matter moving about in a vacuum. We can rub our hands together and find that they become quite warm after a time. When matter changes states or is moving and/or colliding with one another, the standard rules of thermodynamics apply, and generally, the calculations performed are measures of increases and decreases in heat, dissipation rate over time and various momentum changes from speed. In the non-*quantum mechanics* world using the standard model of physics, the *fluctuation dissipation theorem* (or FDT) helps predict the behavior of such things as light absorption and thermal radiation absorption causing matter to slow in velocity. FDT was discovered by Harry Nyquist in 1928 [19] and later proven conclusively by Herbert Callen and Theodore A. Welton in 1951. [20] Unfortunately, the use of FDT equations simply doesn't apply to the *quantum mechanics* nature of black holes and universal expansion.

The number of additional stars born in the cosmos always increases and those resulting stars then eventually die (increasing the chances that more stars will eventually become black holes in the expanding universe) and the fabric they can necessarily exist in continues to grow exponentially. Therefore, experiments observing (or which have been performed initially with radio telescopes) shifts in the movement and spectrum of light over time all produce the same thrilling result, regardless of the nomenclature or formulas used...that the universe is not only expanding, but is doing so at an increasing rate.

As mentioned in the introduction of the book, several experiments have been performed that measure the overall shift in the universe over time, from our perspective on Earth. The identical locations are compared over time, revealing some "expansion" or "stretch" that's known in the astrophysics world as redshift. The question about why a shift in the universes position from our viewpoint on Earth should necessarily be red you've already learned without realizing it. In the electromagnetic spectrum, red represents an increase in wavelength. Blue represents a decrease in wavelength and is typical of an object moving toward the observer; conversely red indicates an object is moving away from the observer. Accordingly, these two shifts that are used in measuring such expansion or contraction are dubbed redshift and blueshift. A person unfamiliar with astrophysics can note the colors prominent in each area or object and then examine the exact same location on a photo series taken years earlier or later and discern (generally) whether the area or object(s) in question are moving inwards, or expanding outwards. The use of spectroscopic redshift observations in contemporary physics is possible because of earlier work in this area in the early twentieth century. While the first recorded use of the written term was unhyphenated as "red-shift" in 1908 by Walter S. Adams [21] it wasn't until 1912 that Vesto Slipher found that all the spiral galaxies he'd examined appeared to have redshifts in them. [22]

More contemporary work in measuring redshifts predominantly has indicated that the universe is not only expanding (due to the "red" high frequency) but it's expanding at an increasing rate over time. These studies been aided considerably with both optical technology (such as the Hubble telescope) and with the many radio telescopes situated throughout the world. The image on the following page shows a recent redshift observation (using the Hubble telescope) against those performed earlier in time across the same regions; the difference is noticeable and is the result of extragalactic jets at work.

Hubble Space Telescope Ultra Deep Field Image, 2012

Courtesy: NASA and ESA

Examining the snapshots in the 2012 Hubble telescope images reveal that the "z-score" or statistical measurement delta begins with a value of 8.6 and ends with an 11.9 value. The numerical difference of 3.3 is the amount of expansion that's occurred for the duration of the time series. To better understand the mapping of a certain point in the cosmos' electromagnetic radiation spectrum by colors, the following chart depicts the spectrums overall scale, from blueshift and up to redshift. [23]

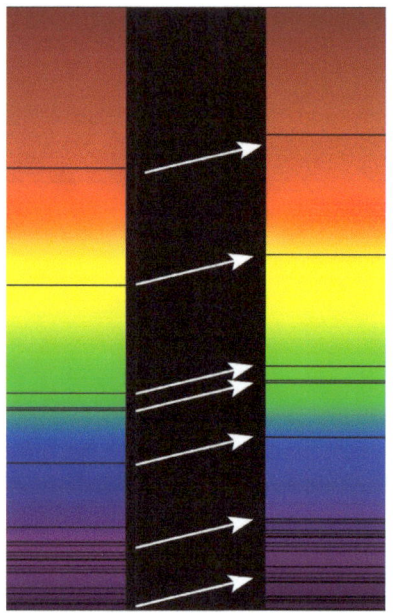

These studies cannot necessarily chart the frequency shifts of the entire cosmos, but they do provide a general indication of how fast things appear to be expanding from our position as observers on Earth. One can envision this as measuring how fast a raft is traveling down a swift moving river from a fixed spot on the shore. The river is longer than we can see (typically) from each direction and so the raft may increase in speed, slow down, or even become stopped by getting caught on a floating tree. The river will continue to flow, the speed of the water may fluctuate and the landscape may change dramatically outside the observer's field of view; there could be a peaceful lake, rapids, a waterfall or even a giant dam lurking out-of-sight, but still a part of the same river where the observer is standing. The complexity of time dilation (for the observer on the riverbank) is also a variable that may not be predictable; it may be impossible for the observer to know that the river is flowing into the future outside of their view, for example. As such, all the observer can do to understand the dynamics of the rivers flow is measure it at as many places as possible and compare them over time. In our illustration, the universe is the river and we cannot simply measure it from its starting and ending locations. We can only compare the speed and temperature of the water in the river from the places we can see from one fixed location. While we may get excellent data about the hydrology in the places we can see to measure, it's difficult to predict with any confidence that those same conditions might then exist outside our immediate view of the river.

The *Eastman constant*, which was discussed in chapter three, is less than perfect and can only use the number of black holes we've discovered from Earth as its dataset, however, it helps to provide a more circular or 360° view of the universe expanding from our position on Earth. This opposes the redshift experiments that are typically constrained to one field of view, such as galaxies in one direction from Earth. The more solar mass that a black hole has accumulated suggests that not only that its gravitation has increased, but that its consumed more information and has extragalactic jets moving outwards from them at a higher velocity. Using the two measurements in conjunction allows for more confidence of the true expansion rate of the universe from our position on Earth. Neither technique is especially useful for accounting for time dilation factors either; as we're limited to that which we can observe from Earth, but, they offer compelling evidence as to how and why the universe is expanding at an increasing rate. The number of observable black holes may not be in a convenient distribution to account for the entire galaxy, however, there's many of them spread very far apart.

The final element of the "radiation problem" extended to *Hawking radiation* (via the extragalactic jets that emanate from black holes) is that they have no dissipation rate which eventually will cause them to slow their "stretching" activity and cause them to then simply exist statically in the region of space their velocity normalizes. It. As the number of black holes increase, the dissipation rate of extragalactic jets becomes "stuck" in a race condition; the number of new, fast moving jets stretching the fabric of space may outnumber those who've physically dissipated. It's helpful to consider the half-life of radioactive isotopes in a nuclear reactor; it may take 20,000 years for the radioactivity to decrease just halfway, however, if you add an exponential number of new isotopes to the storage pond every year, you'll soon reach the same point that extragalactic jets face; there's simply more deadly radioactivity entering the storage pond than time can ever naturally reduce. In this example, the available room in the storage pond would quickly be gone and new ponds would need to be constructed continually (at a faster rate over time) to just keep up with the constantly incoming spent isotopes that require storage.

Let's return once more to the thermodynamics of an extragalactic jet that's been created by gravitational absorption, for a moment. Light absorption necessarily causes light energy to change into heat. The corresponding alternation of this is thermal radiation; it's turned heat energy into light energy. Thermal radiation is easiest to imagine as a glowing (or throbbing) instance of a color in the electromagnetic spectrum, for example. Furthermore, the more an object absorbs light in a vacuum, the more thermal radiation it then emits. This phenomenon is known as *Kirchhoff's law* of thermal radiation [24] and is helpful for understanding how to calculate the dissipation of thermal radiation, explaining its various characteristics and variance. In following *Kirchhoff's law*, it becomes apparent that this path is not a successful one for *Hawking radiation* either; despite the amount of thermal radiation inherent to information in extragalactic jet, the velocity its traveling at exceeds the speed of light and will never slow down enough to use thermal radiation as a basis of validating the "evaporative" effect inherent in *Hawking radiation.*

In closing, it's likely not possible that *Hawking radiation* exists in black holes; the information which has been consumed by a black hole is stretching the fabric of space-time itself and will never, ever stop or slow down. It's also entirely possible that *Hawking radiation* has nothing to do with such consumed information; but only if its existence is found to be not commensurate with *Cherenkov radiation*, and, that the emitted radiation is following a non-quantum model that's never originally exceeded the speed of light. The assertions concerning copies of information being made upon entry of a singularity is inherently false; unless one accepts that its velocity has exceeded the speed of light, and, that as an observer it's passing directly past us on Earth at that moment. As recalled in the previous chapter, the observer would see one version of the information going in each direction.

I hope the discoveries unveiled from my many thought experiments contained in this work may prove useful to the casual reader as a solid explanation for a difficult phenomenon, while serving as a foundation for further debate and exploration by those in academia and quantum physics.

APPENDIX 1

KNOWN SUPERMASSIVE BLACK HOLES [25]

Name	Distance from Earth (LY)	Solar Masses
3C273	2,000,000,000	1,000,000,000
ARP 151	300,000,000	13,000,000
IC 1459	100,000,000	1,000,000,000
M104	32,000,000	660,000,000
M105	38,000,000	130,000,000
M106	24,000,000	31,000,000
M31	2,500,000	30,000,000
M32	2,400,000	3,000,000
M51	27,000,000	1,000,000
M60	51,000,000	4,500,000,000
M77	55,000,000	15,000,000
M81	12,000,000	7,000,000
M84	50,000,000	1,500,000,000
M87	50,000,000	6,600,000,000
Milky Way	27,000	4,100.000
NGC 1023	33,000,000	50,000,000
NGC 1194	170,000,000	65,000,000
NGC 1277	220,000,000	17,000,000,000
NGC 1365	56,000,000	2,000,000

Name	Distance from Earth (LY)	Solar Masses
A0620-00	3,050	7
NGC 2778	76,000,000	20,000,000
NGC 2787	24,000,000	40,000,000
NGC 2960	230,000,000	11,500,000
NGC 3115	32,000,000	900,000,000
NGC 3245	68,000,000	200,000,000
NGC 3377	33,000,000	70,000,000
NGC 3384	38,000,000	16,000,000
NGC 3393	165,000,000	31,000,000
NGC 3516	120,000,000	23,000,000
NGC 3585	70,000,000	340,000,000
NGC 3607	65,000,000	125,000,000
NGC 3608	75,000,000	210,000,000
NGC 3783	130,000,000	9,000,000
NGC 3842	320,000,000	9,700,000,000
NGC 3945	65,000,000	9,000,000
NGC 3998	45,000,000	570,000,000
NGC 4026	50,000,000	210,000,000
NGC 4061	325,000,000	4,500,000,000
NGC 4151	43,000,000	45,000,000

Name	Distance from Earth (LY)	Solar Masses
NGC 4178	55,000,000	200,000
NGC 4253	170,000,000	20,000,000
NGC 4261	100,000,000	800,000,000
NGC 4335	215,000,000	100,000,000
NGC 4342	75,000,000	300,000,000
NGC 4388	62,000,000	8,500,000
NGC 4395	14,000,000	360,000
NGC 4473	50,000,000	100,000,000
NGC 4486b	55,000,000	500,000,000
NGC 4697	40,000,000	175,000,000
NGC 4889	335,000,000	16,000,000,000
NGC 5576	90,000,000	180,000,000
NGC 6240	320,000,000	1,000,000,000
NGC 7052	190,000,000	330,000,000
NGC 7457	43,000,000	3,200,000
NGC 821	100,000,000	75,000,000
RX J1242-11	700,000,000	100,000,000
SDSS J0927+2943	6,500,000,000	600,000,000
ULAS J1120-0641	13,000,000,000	2,000,000,000
123	1,000,000	500,000,000

APPENDIX 2

KNOWN INTERMEDIATE BLACK HOLES

Name	Distance from Earth (LY)	Solar Masses
ESO 243-49 HLX-1	290,000,000	20,000
G1	2,300,000	19,000
M15	34,000	2,000
M74	32,000,000	10,000
M82 X-1	11,000,000	1,000
Omega Centauri	17,000	40,000

M74, as viewed from the Gemini Observatory in Hawaii

APPENDIX 3

KNOWN SMALL BLACK HOLES

Name	Distance from Earth (LY)	Solar Masses
A0620-00	3,050	7
Cygnus X-1	8,000	15
GRO J0422+32	8,100	7
GRO J1655-40	8,000	7
GRS 1009-45	13,000	8.5
GRS 1915+105	8,100	14
GS 1354-64	85,000	9
GS 2000+25	8,800	7
GU Muscae	16,500	7
GX 339-4	27,000	7
H1705-25	27,000	6
IC 10 X-1	2,400.000	23
IGR J17497-2821	27,000	10
LMC X-1	165,000	10.9
LMC X-3	165,000	9
M33 X-7	2,700,000	15.7
MAXI J1659-152	28,000	7
NGC 300 X-1	6,000,000	15
SS433	16,000	16

Name	Distance from Earth (LY)	Solar Masses
V4641 Sagittarii	32,000	6
XTE J1118+480	6,000	7
XTE J1550-564	17,000	12.5
XTE J1650-500	26,000	3.8

APPENDIX 4

HISTORY OF BLACK HOLE PHYSICS

1640 — Ismaël Bullialdus suggests an inverse-square gravitational force law.

1676 — Ole Rømer proves that light appears to have a finite speed.

1687 — Isaac Newton publishes the law of universal gravitation in a three-volume book called *The Prencipa*.

1758 — Rudjer Josip Boscovich develops the *Theory of Forces*, in which gravity can be repulsive across small distances. Boscovich noted that strange classical bodies, such as white holes, could exist, which wouldn't allow other bodies to ever reach their surfaces.

1783 — John Michell theorizes that special bodies may exist; which may prevent light itself from escaping, further illustrating that a dark star the size of our sun would be only a few miles wide, at best. His theory would later be proven to be correct.

1796 — Pierre-Simon Laplace posits the same conclusion as John Michell; that the largest bodies (black holes) in the universe may be invisible.

1798 — Henry Cavendish measures the gravitational constant, *G*.

1876 — William Kingdon Clifford suggested that the motion of matter might be simply due to changes in the geometry of space.

1909 — Albert Einstein and Marcel Grossmann began

developing a theory that would bind a metric tensor; thus defining space geometry without mass, and, still possessing a source of gravity.

1910 — Hans Reissner and Gunnar Nordström define what's known as a *Reissner-Nordström singularity*.

1915 — Einstein expands general relativity to include the effects of gravity; showing that the larger an object is, the greater the gravitation around it becomes. This allowed the theoretical possibility of a black hole's existence.

1916 — Karl Schwarzschild solves the *Einstein vacuum field equations* for uncharged, spherically symmetrical, non-rotating bodies. Schwarzschild advanced the concept that an extreme amount of mass imploding upon itself creates such intensive gravitation that not even light can escape its pull.

1918 — Hans Reissner and Gunnar Nordström solve the *Einstein–Maxwell field equations* for charged spherically symmetric non-rotating systems.

1918 — Friedrich Kottler achieves the *Schwarzschild radius* solution without the usage of *Einstein vacuum field equations*.

1923 — George David Birkhoff proves that the Schwarzschild space-time geometry is the unique spherically symmetric solution of *Einstein vacuum field equations*.

1931 — Subrahmanyan Chandrasekhar uses special relativity to calculate that a non-rotating body of electron-degenerate matter above a certain mass (at approximately 1.4 solar masses) has no stable solution. This demonstrated that heavier stars would end their lives far differently than conventional stars.

1939 — Robert Oppenheimer and Hartland Snyder calculate the gravitational collapse of a pressure-free, homogeneous fluid sphere as a result of atomic fusion work.

1958 — David Finkelstein theorizes that the *Schwarzschild radius* of black holes is a causality barrier, otherwise known as an event horizon.

1963 — Roy Kerr solves *Einstein vacuum field equations* for uncharged symmetrically rotating systems, deriving the *Kerr metric.*

1963 — Maarten Schmidt researches the quasar *3c273* and his work leads to the theory that all quasars in the universe are subsequently powered by black holes in the center of galaxies.

1964 — Roger Penrose proves that an imploding star can produce a singularity, once it has formed an event horizon.

1965 — Ezra T. Newman, E. Couch, K. Chinnapared, A. Exton, A. Prakash, and Robert Torrance solve *Einstein-Maxwell field equations* for charged rotating systems.

1967 — Werner Israel presented his proof of what became known as the no-hair theorem at *King's College London.*

1967 — John Archibald Wheeler popularizes the term black hole for collapsed stars at a conference in New York.

1968 — Brandon Carter uses *Hamilton–Jacobi theory* to derive first-order equations of motion for charged particles, which are moving in the external areas of a *Kerr-Newman* black hole.

1969 — Roger Penrose discusses the Penrose process for extracting the spin energy from a Kerr black hole.

1969 — Roger Penrose proposes the cosmic censorship hypothesis.

1971 — Identification of *Cygnus X-1* as a black hole is confirmed using combined optical, radio and x-ray data.

1972 — Stephen Hawking proves that the area of a classical black hole's event horizon cannot ever decrease.

1972 — James Bardeen, Brandon Carter, and Stephen Hawking propose four laws of black hole mechanics that agree with the laws of thermodynamics.

1972 — Jacob Bekenstein suggests that black holes have entropy that's proportional to their surface area, which is due to information loss.

1974 — Stephen Hawking applies quantum field theory to black hole space-time and shows that black holes will radiate particles with a blackbody spectrum, known as *Hawking radiation*, which causes evaporation over time.

1989 — The "Black Hole War" begins at a conference in San Francisco with Leonard Susskind, Gerard 't Hooft and others; challenging Hawking's claim that information consumed by black holes is lost forever by evaporation.

1989 — Formal confirmation of *V404 Cygni* as a black hole is made.

1993 — The *CGHS theory* is debated at a conference in Santa Barbara, California, with the *information paradox* contained in *Hawking radiation* being voted against by most the physicist's present; effectively disagreeing with information loss as presented by Hawking and suggesting a principle called *Black Hole Complementarity*. This principal was written in an academic paper by Don Page and advocated by Leonard Susskind and others.

2002 — The *Max Planck Institute for Extraterrestrial Physics* presents evidence to hypothesize that *Sagittarius A*★ is a supermassive black hole in the Milky Way's center.

2002 — NASA's Chandra X-ray Observatory identifies dual galactic black holes in merging galaxies of *NGC 6240*.

2004 — Further observations by *UCLA* researchers present stronger evidence supporting the possibility that *Sagittarius A*★ is a black hole.

2007 — Stephen Hawking formally concedes and thus ends the "Black Hole War" in a matter of sorts.

2012 — The first visual proof of the existence of black holes occurs as Suvi Gezari's team at *Johns Hopkins University* uses the Hawaiian telescope *PAN-STARRS* to publish images of a supermassive black hole located approximately 2.7 million light-years away from Earth. This black hole also appears to be consuming a red giant.

ABOUT THE AUTHOR

Darren completed undergraduate and graduate degree programs in political science at *California State University, Fullerton.* He lives in Silicon Valley and works as a technologist and writer.

Of particular interest to readers of this book, the author was kicked out of AP Physics class in his American high school during the very first week. The teacher announced to the classroom that he should, "focus on playing football instead of physics."

Darren was conferred a *Baron of the German Nation* by the Order of Teutonic Knights in 2014 and is active civically in his community.

REFERENCES

[1] Hawking, Stephen. *Black Hole Explosions? Nature.* Volume 248, 1974.

[2] Lockyer, Norman, Sir. *Nature.* Volume 106. *Macmillan Journals Limited*, 1921.

[3] Schwarzschild, Karl. *Über das Gravitationsfeld eines Massenpunktes nach der Einsteinschen Theorie. Sitzungsberichte der Deutschen Akademie der Wissenschaften zu Berlin, Klasse für Mathematik, Physik, und Technik*, 1916.

[4] The users "MissMJ" and "Ylebru" created this image.
https://commons.wikimedia.org/wiki/File:World_line2-it.svg

[6] Image courtesy of the *Caltech* & *Cornell University* SXS Collaboration.
http://astro.cornell.edu/sxs-collaboration.html

[7] Second Quantum Gravity Conference. Rutherford-Appleton Laboratory, *Oxford University.* February 1974.

[8] Liouville, Joseph. *Journal de Mathématiques Pures et Appliquées*, 1846.

[9] Susskind, Leonard. *The Black Hole War: My Battle with Stephen Hawking to Make the World Safe for Quantum Mechanics.* Back Bay Books, 2009.

[10] Bekenstein, Jacob. *Black Holes and Everyday Physics. Gravity Research Foundation*, 1981.

[11] Flamm, Ludwig. *Beiträge zur Einsteinschen Gravitationstheorie. Physikalische Zeitscrift*, 1916.

[12] Einstein, Albert. *On the Electrodynamics of Moving Bodies. Annalen der Physik* 17:1891, 1905.

[13] Einstein, Albert. Covariance *Properties of the Field Equations of the Theory of Gravitation Based on the Generalized Theory of Relativity. Zeitschrift für Mathematik und Physik*, 63, 215–225, 1915.

[14] Feinberg, Gerald. *Possibility of Faster-Than-Light Particles. Physical Review* 159-5, 1967.

[15] Cherenkov, Pavel. *Visible Emission of Clean Liquids by Action of γ Radiation. Doklady Akademii Nauk SSSR* 2: 451, 1934.

[16] Hamilton, William Rowan. *On a General Method of Expressing the Paths of Light, and of the Planets, by the Coefficients of a Characteristic Function. Dublin University Review*, 1, 1833.

[17] Lipunov, Vladimir. *Tachyon Motion in a Black Hole Gravitational Field. Astrometriia i Astrofizika*, 35, 1978.

[18] *CERN* Press Office. *OPERA Experiment Reports Anomaly in Flight Time of Neutrinos from CERN to Gran Sasso.* September 23, 2011. http://press.web.cern.ch/press-releases/2011/09/opera-experiment-reports-anomaly-flight-time-neutrinos-cern-gran-sasso

[19] Chandler, David. *Introduction to Modern Statistical Mechanics. Oxford University Press*, 1987.

[20] Caleen, Herbert. Welton, Theodore, A. *Irreversibility and Generalized Noise. Physical Review.* 83-34, 1951.

[21] Adams, Walter, S. *Preliminary Catalogue of Lines Affected in Sunspots. Carnegie Institute of Washington.* 22 1–21, 1908.

[22] Slipher, Vesto. *The Radial Velocity of the Andromeda Nebulae. Lowell Observatory Bulletin.* 1:256, 1912.

[23] Image created by Georg Wiora in February 2011.

[24] Kirchhoff, Gustav. *Ueber das Verhältnis zwischen dem Emissionsvermögen und dem Absorptionsvermögen der Körper für Wärme and Licht. Annalen der Physik und Chemie.* 109, 275–301, 1860.

[25] The number of black holes contained here aren't meant to be an exhaustive list of every black hole detectable from Earth; this information changes over time as new discoveries and measurements are made.